零度城市研究系列　李涵　金秋野　著　

HUTONG MUSHROOM
Urban Studies Degree Zero Series
LI HAN　JIN QIUYE

汉英对照

中国建筑工业出版社

目 录
Contents

不要问我什么是城市研究	5
Don't Ask Me What Urban Studies Is	14
胡同蘑菇模型 Models of Hutong Mushroom	27
胡同蘑菇图画 Drawings of Hutong Mushroom	99
胡同一课	151
A Lesson from Hutong	158
研究团队 Research Team	168
作者简介 About the Authors	172

李涵

不要问我什么是城市研究

1. 原型与表象

这几年,我们一帮子落魄的建筑师"野武士"[1]们都在修炼城市研究。不知何种原因,大家在欣赏城市的口味上表现出了高度的一致。比如,王硕[2]特喜欢在南城大红门一带天天清仓的服装市场淘货;朱起鹏[3]每晚都要到鼓楼后边一个小破庙观(宏恩观)打两局美式落袋,再到旁边的饭馆吃碗卤煮当晚餐;学历最高的清华在读博士徐腾[4]跋山涉水到河北易县朝拜彩钢板搭出的"他奶奶的庙"。而我则把周末夜晚的美好时光挥霍到三里屯脏街的酒吧里,在震耳欲聋的噪声里喝着杂牌兑水威士忌。我们热爱这些脏乱差的城市角落,并且一本正经地对这些角落进行城市研究。尽管问我什么是城市研究,就像问我什么是宇宙一样,我可以啰唆一小时,但其实我困惑得一塌糊涂。

正当我们撸起袖子,在各自钟情的城市角落里苦练城市研究内功时,噩耗频传。王硕的大红门率先关门,迁往河北深处。朱起鹏的宏恩观随即遭整顿,变成了一个他无法忍受的古色古香的茶楼。不久前,三里屯脏街也在集中整治开墙打洞的运动中变得干干净净。严酷的现实让我不得不认定:功,练歪了!

为此,王硕组织了"北京杂种"论坛,让我们这帮失魂落魄的野武士相互安慰。

我不怀好意地问王硕:"你这大红门怎么办啊?咋研究研究着就没了呢?"王硕不以为然:"我们所做的是要找到它们内在的运行机制,并从这种运行机制中,提取出它的空间原型,然后将这种知识的生产,这种对空间原型的寻找、探索,最终转化成空间的生产。"[5] 按照王硕的观点,城市研究就是要透过纷繁的城市表象看到其核心本质,提取原型就是要把具体事物抽象化,演变成知识,进入流通环节,从而用于其他空间的生产。

对如此强有力的方法论,我的疑问来源于我个人的经验。我第一眼看到三里屯脏街42号楼,就一见钟情,要把它作为研究对象。一开始,我同样是抱着拨开浮云看本质的思路去研究,但我不能骗自己,我不用拨浮云就知道它的本质原型是:居民楼+底商。这个原型可以有各种变体,变好了是建外SOHO,变不好就成了银河SOHO(本人观点,SOHO跟居民楼类似,都是独立产权的小空间聚集。尽管建外SOHO相对简陋,但由于规模、位置等因素,人气很旺,从城市角度出发是成功的。而银河SOHO尽管是扎哈·哈迪德(Zaha Hadid)设计的超炫建筑,但人迹罕至,拍片不用清场,成为著名影视外景地)。这个抽象的原型,或者说这个"知识"对我来说并不新鲜,但我还是被42号楼深深地吸引。吸引我的,毫无疑问是具体的表象:一个浓妆艳抹的年轻女子坐在白色塑钢窗包裹的居民楼阳台上,阳台已经变成了麻辣烫店的雅座,阳台下面单元门的入口挂满了霓虹灯,旁边几个外国人在黑影里窃窃私语,再远处是派出所蓝色的灯箱和一堆警车。这些奇妙的组合、这些具体的元素，让我的眼睛停不下来。但一旦回到那个抽象的原型，一切索然无味。当我意识到自己在城市研究上竟然是个反智人士时，我所能做的就是用图画去表现浮华的表象了。

2. 谁是"零度"

王骏阳老师的论文《日常：建筑学的一个"零度"议题》[6]是关于"日常"对建筑学影响的一次全面梳理。我特别感兴趣的是他文中提到的"零度"这个概念。按照王骏阳老师的文章：

> "零度"概念源自罗兰·巴特（Roland Barthes）和他的《写作的零度》……萨特（Jean-Paul Sartre）将文学视为一种"介入"世界（社会和政治）的方式，承载着文学对于"自由"的责任和使命。……巴特"将文学作为一种'迷思'(myth)，其意义首先不在于呈现世界，而是将自己呈现为文学。"

为文学而文学的巴特（零度）与为社会而文学的萨特之间的这场文学之争以巴特的胜利告终。

但文章进入下一段，"零度"概念翻转了：

> 彼得·埃森曼（Peter Eisenman）在1970年代初提出一个"建筑自主性"理论，这个"自我指涉符号"理论多少像是巴特《写作的零度》的形式主义倾向及其"零度意义"的建筑转化。作为对埃森曼"批判性建筑学"的质疑和批判，埃森曼的学生主张"投射性建筑学"，试图通过"多普勒效应"(the Doppler Effect)来说明打破"批判性建筑学"自闭的学科边界，加强建筑学与外部现实世界的互动关系的必要。

> 学科从批判到投射的转换可以表达为冷却的过程，批判性建筑学之所以是热的，就在于它热衷于将自身与常规的、背景化的以及

匿名的建筑生产状况区分开来,并着力刻画这种差异……如果说"冷却"是一种混合的过程,那么"热"则通过区分进行抵抗,而且意味着极度艰难、繁琐、努力、复杂。"冷"是轻松悠闲的。

这里,为建筑而建筑的埃森曼忽然由"零度"变成"热"的了,为社会而建筑的"投射性建筑学"反而成为"冷",趋向"零度"。我后来有机会找王老师请教,老师跟我说,这个"零度"概念仕义中有多次转换,好好体会吧!

我之所以跟"零度"较劲,还是因为困惑于建筑师该如何做城市研究。"零度"的概念让我对城市研究的模糊想法突然清晰起来,我冒昧地篡改一下王老师的文章,来表达我的观点:

"零度"概念源自李涵的《城市研究的零度》……王硕将城市研究视为一种"介入"世界(社会和政治)的方式,承载着研究对于"知识生产"和"空间生产"的责任和使命。……李涵将城市研究作为一种"迷思"(myth),其意义首先不在于知识生产和空间生产,而是将自己呈现为城市作品(它既可以是文字、图像,也可以是视频、模型)本身。

作品意味着具有独立的欣赏价值的完成品。它自成一体,它就是它自己,它就是最终结果,它无需进入其他环节。

按照文学的冷热标准,"零度城市研究"获得胜利。

然而以科学理论评判冷热,又是另一番景象:

李涵在2017年提出一个"城市研究自主性"理论。作为对"自主性"的质疑和批判,王硕主张"投射性城市研究"。他试图通过"多普勒效应"说明打破"城市研究自主性"自闭的学科边界,加强城市研究与外部现实世界的互动关系的必要。从自主到投射的转换可以表达为冷却的过程。自主性城市研究之所以是热的，就在于它热衷于将自身与其他领域区分开来，在自我"批判"和"超越"中自娱自乐。这就是它的"热"，而从"自主"进入现实，致力于与学科外部条件的融合和互动则是"投射性城市研究"倡导的"冷"，如果说"冷却"是一种混合的过程，那么"热"则通过区分进行抵抗，而且意味着极度艰难、繁琐、努力、复杂。"冷"是轻松悠闲的。

按照科学理论的冷热标准，代表"零度"的"投射性城市研究"胜利，冷热反转了。

3. "零度城市研究"实验

当金秋野老师邀请我带一个城市研究课题时，我选择尝试文学标准的"零度城市研究"。

我们有 100 多位研究人员，人数是我们的优势。研究的范围是整个北京，如果研究人员喜欢，范围可以更大，因为每个研究人员只需挑出一幢"自然长成的建筑"即可。"自然长成"，意思是没有经过专业人员设计的，被肆意使用着的，被尼古拉斯·佩夫斯纳（Nikolaus Pevsner）称为"建（贱）物"，被冢本由晴（Yoshiharu Tsukamoto）称为"坏建筑"，被我称为"草根建筑"的构筑物。当然，相对于每位研究人员的工作，这应该算作建筑研究。但作为总体，100 多个建筑放在一起时，叫"城市研究"未尝不可。

作为"零度城市研究"，我们有绝对偏执的要求。

"零度"对于文学就是文字本身，对于绘画就是画面本身，对

于建筑就是建筑本身。一切从完成品（作品）出发。具体到城市研究，有3种主要完成品（当然还有其他形式）：文字、图纸和模型。"零度城市研究"不关心研究人员如何调研，也不关心他们如何分析。过程没有意义，让各种概念、草图、分析图、表格都歇了吧。我们只关注最终呈现的完成品是否达到独立欣赏的"金线"。[7] 我们只从结果出发，我们也只看结果。

对于文字，我们看文章本身的文采。虽然这不是文学研究，但一部关于城市的优秀小说，例如查尔斯·狄更斯（Charles Dickens）的《雾都孤儿》，难道不是真正优秀的城市研究吗？建筑师同样可以写出优秀的文字，一篇好的论文，例如柯林·罗（Colin Rowe）的《透明性》，难道不是优秀的文学吗？"零度"研究不是按学科分类，而是根据完成品本身的形式设定判断标准。文字就按文学的价值判断，图纸就按绘画的价值判断，绝不能因为这是建筑师关于城市的研究论文，就可以容忍劣质的文字出现。这就是"零度"的严苛，"零度"是容忍的零度！鉴于此，我们从一开始就禁止研究人员生产任何文字形式的研究成果，因为他们的写作经验太有限，很难达到文字作品的金线。

对于图纸，我们看画面本身的格调，看颜色，看构图，看造型，以及看整个画面传达出的质量。1963年，约翰·海杜克（John Hejduk）在康奈尔大学教课，隔壁班的学生不停地跑到他的班里，看上去紧张、担心，甚至有些抑郁，他们抱怨："那家伙疯了，他让我们把一个楼梯间画了100遍，有一根线条不完美，就整张重画。""我们绝对不许使用橡皮，这家伙就只会说'要画得完美'，其他什么也不讲，然后看一眼，就告诉我们重画。"海杜克立刻对这家伙肃然起敬。他后来见到了这位偏执的老师：斯坦利·泰格曼（Stanley

Tigerman)。[8] 讲这个故事就是要说,泰格曼的要求就是"零度"的要求,对于图来说,画什么是第二层次的,画得好不好才是首要的。我们宁愿要一个画得完美的私搭乱建的阁楼,也不要一个画得草率,但听起来似乎很酷的概念。

对于模型,我们看模型本身的品质。对于年轻的研究人员,模型是"零度城市研究"最好的领域。只要认真、细致、勤勉,加上一些基本的技巧,模型就可以不断地吸收研究人员的能量,达到独立欣赏的金线。这些模型绝不是工作模型,不是研究用的,也不是分析用的,它就是作为模型自身而存在。就像家本由晴说的,画轴测图是为了爱上那些被画的建筑。做这些模型也是让研究人员爱上那些草根建筑,进而把这份感情迁移到模型中,从而让模型获得作品的力量。

这场"零度城市研究"实验的成果是超过作品金线的几十个模型和十几张图纸。这里面或许包含几种空间原型、若干构造体系,类型学的研究、知识的生产可由此展开。"科学标准""零度"研究似乎以此为基础可以开始了。但这样理解就错了,两种"零度"研究,其实从一开始就分道扬镳了:分析原型、研究构造、生产知识,根本用不着画得如此完美的图纸,做得如此精致的模型。研究人员用不着爱上它们,只需冷静、清晰、有逻辑地分析和说明即可。这些模型和图纸不是为了知识生产而存在的,它们就为了自身而存在,为了独立欣赏而存在,为了作品生产而存在,它们放在一起不是用来分类的,找原型的,而是将自己呈现为关于北京的作品。

这还是城市研究吗?从起初的研究对象上就歪了,到后来的研究方法上更是走火入魔了。所以请不要问我什么是城市研究!

注释：

1. 日本建筑师桢文彦将伊东丰雄、安藤忠雄、石井和紘这一批 1940 年代出生的建筑师称为"和平时代的野武士"。野武士是日本战国后期一群带刀的落魄户，在各地主番臣间流浪，一有战事，即被招募。野武士的报酬很低，给饭吃就可以，甚至制服都可以省略。

2. 王硕，建筑师，META- 工作室创始合伙人。

3. 朱起鹏，建筑师，中国古迹遗址保护协会（ICOMOS CHINA）会员，神奇建筑研究室合伙人 / 设计主持人。

4. 徐腾，不正经历史研究所所长。

5. 出自王硕在"听到讲坛"上的演讲《城市野生状态下的空间生产》。

6. 原文刊载于《建筑学报》2016 年 10 期。

7. "金线"的概念出自作家冯唐给韩寒的一封信："文学的标准的确很难量化，但是文学的确有一条金线，一部作品达到了就是达到了，没达到就是没达到，对于门外人，若隐若现，对于明眼人，一清二楚，洞若观火。"

8. John Hejduk, "Afterword", in Stanley Tigerman Buildings and Projects (Rizzoli International Publications, INC. 1989).

Li Han

Don't Ask Me
What Urban Studies Is

1. Prototype & Phenomenon

In recent years, a bunch of us, who regarded ourselves as "wild samurai architects"[1], had been cultivating in the field of urban studies. For an unknown reason, we shared a common taste that was highly consistent on our urban interests. For instance, Wang Shuo[2] was fond of bargain hunting in the cloth bazaar near the Great Red Gate (Dahongmen) in the south of Beijing. Zhu Qipeng[3] would attend an American pool every night in a shabby Taoist temple (Hong En Guan) hidden somewhere behind the Drum Tower (Gulou) with a bowl of gut hotchpotch (luzhu) for dinner afterwards. Xu Teng[4], a PhD candidate in Tsinghua University, had the shiniest education among us. He drove all the way to Yi county in Hebei province and went on a pilgrimage to "the Temple of One's Grandma" made of tatty prefabricated shelters. Me? I spent almost all of my lovely weekend nights in the nasty, crowded pubs located within the chaotic Sanlitun, having cheap, watered down, unbranded whiskey. We all loved those casual corners of the city. We all did serious urban studies on them. Although I could respond to the question "what is urban studies" with a whole bunch of nonsense for an hour just like answering "what is the universe", I am actually extremely confused.

While we were rolling up our sleeves and getting ready for a battle within our own corners of the city, bad news was all over the place. Firstly, the Great Red Gate bazaar (Dahongmen) of Wang Shuo's was closed and moved out of Beijing deep into Hebei. Then Zhu Qipeng's shabby Taoist temple (Hong En Guan) was forcibly renovated into an ordinary tea house that he could never put up with. Recently, the chaotic Sanlitun was totally cleaned up in the official "remediation" movement, in which boutiques were closed and walls aside were built. These harsh realities forced me to believe that our cultivation was crooked.

Wang Shuo organized a forum called "the Beijing Bastards" for all of us. We, the wild samurais, lost our wits. We could only provide some comfort for each other.

Meanwhile I asked Wang Shuo with an evil smile: "What are you going to do about your beloved Dahongmen bazaar? How come it disappeared out of a sudden while you were researching it?" Wang Shuo didn't care much about me being mean: "What we are about to do is to think about its internal mechanism. From the operating mechanism, we could extract its spatial prototype. The knowledge and the process of pursuing a spatial prototype would eventually be transformed into spatial production."[5] According to Wang Shuo, urban studies would be reaching the core essence of city through all the complex urban phenomena. An extracted prototype would be the abstraction of a specific matter. It could be transformed into knowledge for academic circulation.

Therefore it could be used in the production of other spaces.

Such a strong methodology. However, I question it from my own experience. The first time that I saw the Building 42, located on the well-known Sanlitun Dirty Street, I fell in love with it instantly. I told myself that I would study it. At the beginning, I was also carrying the methodology of finding an essence behind everything. But I kept failing at lying to myself that I had simply known its essential prototype from the very first day: ground floor commercial space plus residential space above. This prototype could have various of mutations. A benign mutation would be Jianwai SOHO. A malignant mutation would be Galaxy SOHO (both of the SOHOs are projects by a local real estate company in Beijing. In my opinion, they are similar to residential buildings. Although Jianwai SOHO is rather simple and crude, its scale and location cause a boom in foot traffic. It is a success in urban perspective. Even though Galaxy SOHO has a cool appearance by Zaha Hadid, it has been relatively desolate. Movie crews love it since they don't even bother to do site-clearing. Now it is pretty famous as a filming site). This abstract prototype, or the knowledge of it, was nothing new to me at all. But I was still fascinated by the Building 42. Undoubtedly, those specific phenomena were attracting me instead of anything else. A young woman with heavy makeup sitting on the balcony covered with a white plastic steel window. The balcony happened to be the lounge of a street stall. Beneath the balcony there is a unit entrance. Neon lights were hanging messily all over it. Several foreigners were whispering in the shadow, while police lights and cars were not so far from them. All these magical combinations and specific elements kept my eyes busy observing. But once my mind jumped back to the abstract prototype, it was completely dull and boring. I surprisingly realized that I had always been an anti-intellectualist in terms of urban studies. All that I could do about it, was to represent the superficial phenomena through my drawings.

2. Who is "Degree Zero"?

"The Everyday: A Degree Zero Agenda for Architecture"[6], a paper written by Professor Wang Junyang, was an overall inspection of the impact that "Everyday Life" has on architecture. I was particularly interested in the concept of "Degree Zero" that Professor Wang mentioned in this paper, which was interpreted as:

The concept of "Degree Zero" originated from Roland Barthes and his "Writing Degree Zero" (Le Degré zéro de l'écriture) … Jean-Paul Sartre regarded literature as an "intervention" into the world (socially and politically) which bears the responsibility and mission of "liberty" … Barthes "regarded literature as 'myth'. The significance of which was not to present the world but to present itself as literature."

This debate was between Barthes' "literature for literature" and Sartre's "literature for society". Barthes won the debate.

However, when it came to the next paragraph, Professor Wang reversed the concept of "Degree Zero":

In the beginning of 1970s, Peter Eisenman proposed a theory of "the Autonomy of Architecture". This "self-reflexivity" theory somewhat resembled Barthes' tendency towards formalism and was his "Degree Zero" concept in architecture.

As a critique of Eisenman's "Critical Architecture", his students proposed "Projective Architecture", trying to explain the necessity of breaking the self-closed disciplinary boundaries of "Critical Architecture" and enhancing the interaction between architecture and the real world outside, via the concept of "the Doppler Effect".

The transformation from criticism to projection of a discipline can be articulated as a cool-down process. Critical Architecture is "hot" because of its enthusiasm in differentiating itself from normal, backgrounded and anonymous architectural production, and endeavoring to highlight such differences… If we say "cool-down" is a process of mixing up, then "hot" resists through differentiation, which means extreme hardness, intricacies, endeavor and complexity, while "cold" is more casual and leisurely.

Peter Eisenman's "architecture for architecture" was defined as "hot" here, instead of "Degree Zero". Conversely, "Projective Architecture", which claimed "architecture for society" became "cold" and tended to be "Degree Zero". When I had a chance to consult with Professor Wang, he told me that the concept of "Degree Zero" reversed several times in this article, and that I should think twice about it.

I was struggling with the definition of "Degree Zero", because I was still confused about how architects should conduct urban studies. The concept of "Degree Zero" gave me a sudden enlightenment on urban studies. Therefore I took the initiative to modify Professor Wang's article in order to present my views:

The concept of "Degree Zero" originated from Li Han's "Urban Studies Degree Zero" … Wang Shuo regards urban studies as an "intervention" into the world (socially and politically) which bears the responsibility and mission of "knowledge production" and "space production" …Li Han regards urban studies as a "myth". The significance of which is not to produce knowledge or space, but to present themselves as works about the city (which could be writings, images, videos, or models).

When a piece of work is fully realized, it will have its independent aesthetic value. It is complete on its own. It is what it is. It is the final outcome. It does not require any other procedures.

Referring to the criterion of "cold" and "hot" in literature, "Urban Studies Degree Zero" makes more sense.

However, if we compare "cold" and "hot" in terms of scientific theories, it would be another story:

In 2017, Li Han proposed a theory of "the Autonomy of Urban Studies". As a critique of "autonomy", Wang Shuo proposed "Projective Urban Studies", trying to explain the necessity of breaking the self-closed disciplinary boundaries of "Autonomous Urban Studies" and enhancing the interaction between urban studies and the real world outside, via the concept of "the Doppler Effect". The transformation from autonomous to projective urban studies can be articulated as a cool-down process. Autonomous Urban Studies is "hot" because of its enthusiasm in differentiating itself from other fields, and entertaining itself by "criticizing" and "surpassing" itself. This is so-called "hot". However, moving from "autonomy" to reality and endeavoring in the interaction and mergence with external conditions is the "cold" advocated by "Projective Urban Studies". If we say "cool-down" is a process of mixing up, then "hot" resists through differentiation, which means extreme hardness, intricacies, endeavor and complexity, while "cold" is more casual and leisurely.

According to the criterion of "cold" and "hot" in scientific theories, "Projective Urban Studies" would represent "Degree Zero". Thus, the "cold" and "hot" have been reversed.

3. Experiment of Urban Studies Degree Zero

When Professor Jin Qiuye invited me to lead an urban studies project, I chose to experiment with "Urban Studies Degree Zero" under the standard of literature.

To our advantage, we had more than one hundred researchers. The range of the study was the entire city of Beijing. The range could have been even larger if the researchers preferred, because each of them was only required to choose one piece of "naturally grown architecture". The "naturally grown architecture" is defined as what has not been designed by professionals but has been used freely. It is defined as "building" (inferior architecture) by Nikolaus Pevsner, as "Da-me Architecture" by Yoshiharu Tsukamoto, and as "grassroots architecture" by me. Of course, to each researcher, it was an architectural study. However, when more than 100 buildings are collected together, we can refer to it as urban studies.

To "Urban Studies Degree Zero", we have absolutely intolerant standards.

To literature, "Degree Zero" is the writing itself. To drawing, it is the image itself. To architecture, it is the built building itself. Everything starts from the outcome (the finished work). In urban studies, there are three major outcome forms: writing, drawing and model. "Urban Studies Degree Zero" does not care about how the researchers carry out the research or how they analyze it. The process is meaningless. Let's drop all the concepts, sketches, diagrams, and charts. We would only focus on whether the final presentation of the outcome could reach the "golden line"[7] if viewed independently. We only start with the outcome, and we only care about the outcome.

In writings, we look for the literary grace. A great novel about city might not be a study on literature. For instance, Charles Dickens' *Oliver Twist* is a real prominent work in urban studies, isn't it? Architects can also create impressive writings. A good essay, like Colin Rowe's "Transparency", is an excellent piece of literature. "Degree Zero" studies are not categorized by discipline, but the criteria are based on the forms of the finished works. Writings should be judged by their literary value. Drawings should be evaluated on their artistic value. We cannot accept bad writing simply because it is an urban studies essay written by an architect. This is the rigor of "Degree Zero". "Degree Zero" is zero tolerance. Hence, we forbid researchers to produce any research outcome in the form of writing. They can hardly reach the "golden line" of literature with their limited experience in writing.

For drawing, we would examine the style, composition, form and quality of the image as a whole. In 1963, John Hejduk was teaching at Cornell University. Students from the other studio always came to his class, looking nervous, worried, and even depressed. "That guy must be crazy," they complained to Hejduk, "He made us draw a staircase for 100 times. If one line is not perfect, we need to start all over again." Hejduk immediately became awestruck at "that guy". Later, he met this paranoid professor - Stanley Tigerman[8]. Tigerman's standard is the standard of "Degree Zero". For a drawing, its finished quality is more important than its subject. We would rather have a randomly-built but perfectly-drawn attic than a concept that sounds cool but is hastily represented.

Concerning the model, we would evaluate the quality itself. Modeling is the best field in "Urban Studies Degree Zero" for young researchers. With earnestness, care and diligence, models can continuously absorb the essence infused by the researchers and reach the "golden line" when appreciated independently. These models are not working models. The existence of each model is not as a tool for research or analysis, but as a model itself. As Yoshiharu Tsukamoto said, he was creating axonometric drawings in order to fall in love with the drawn architecture. Researchers should be making models for the reason of loving the grassroots architecture, further bringing their emotions into the models and empowering the models as finished works.

The dozens of models and drawings that exceed the "golden line" were the achievements of this "Urban Studies Degree Zero" experiment. These models and drawings might contain several spatial prototypes and construction systems. As a result, the study of typology and knowledge production could begin, and then serve as the basis for "Degree Zero" studies under "the standard of science". However, such an understanding is wrong. These two directions of "Degree Zero" studies have split from the very beginning: analyzing prototypes, studying construction,

and producing knowledge do not require such perfect drawings and delicate models. The researchers do not need to love the subjects. They only need calm and clear introduction and rational analysis. The existence of these models and drawings is not for producing knowledge, but for their own sake, for independent appreciation, and for producing works. They are put together not to be categorized or studied as prototypes, but to be presented as a work about Beijing.

Was this still urban studies? From the beginning, the choice of study subjects was off-track, while the later study method went even more awry. So, please do not ask me what urban studies is.

Notes:

1. Architect Fumihiko Maki called the Japanese architects born in the 1940s "the wild samurais in a peaceful era", including Toyo Ito, Tadao Ando and Kazuhiro Ishii ("The wild samurais" usually referred to the poor warriors with knives during the late Sengoku period of Japan. They were only hired for wars sporadically. Whenever a war was over, they would be homeless and looking for another landlord to fight for. Their income was low. They were satisfied with basic survival, and didn't even dress properly).

2. Wang Shuo, architect, founding partner of META-Project.

3. Zhu Qipeng, architect, ICOMOS CHINA Member / Partner & Chief Designer of Wonder Architects.

4. Xu Teng, founder of Buzhengjing Lishi Yanjiusuo (Research Institution of Not Serious History).

5. Extracted from "Space Production in Urban Wildlife", a speech

by Wang Shuo at Tingdao Forum.

6. Wang Junyang, The Everyday: "A Degree Zero Agenda for Architecture", *Architecture Journal*, 2016-10.

7. The concept of "golden line" comes from writer Feng Tang's letter to writer Han Han: "The standard of literature is very difficult to quantify. However, literature indeed has a clear golden line. If a work reaches it, it does. If it does not reach it, it does not. For outsiders, it seems looming. But for an insider, it is crystal clear."

8. John Hejduk, afterword, *Stanley Tigerman Buildings and Projects* (Rizzoli International Publications, INC. 1989).

胡同蘑菇模型
Models of
Hutong Mushroom

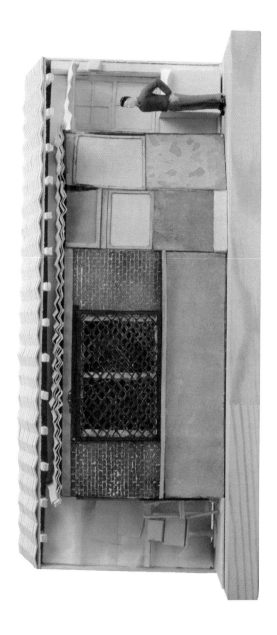

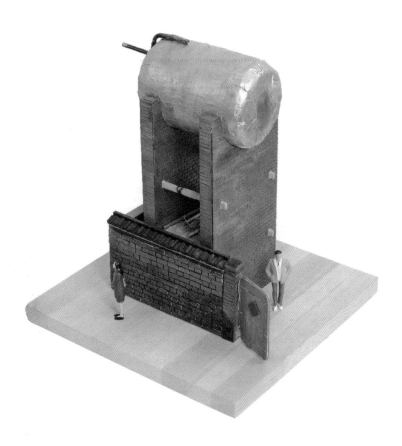

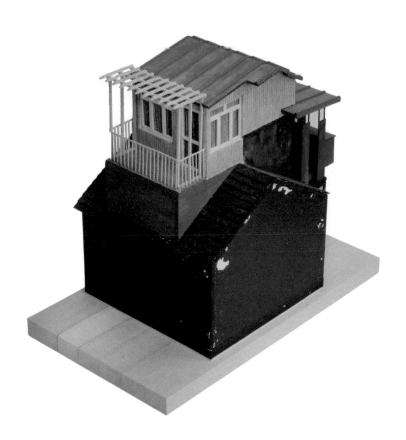

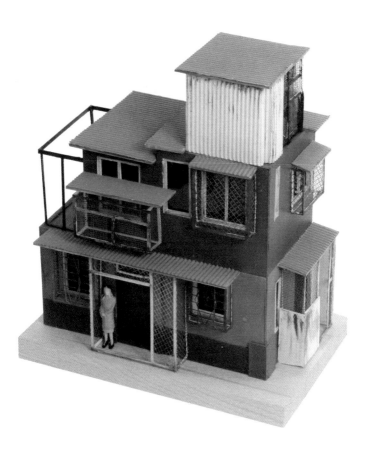

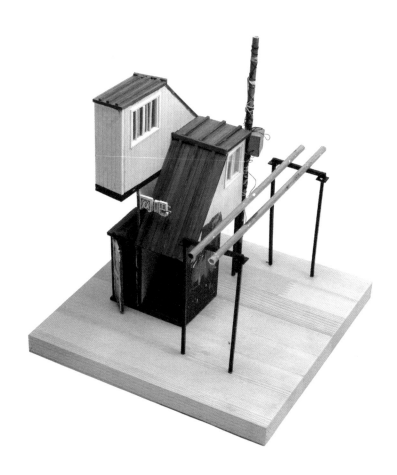

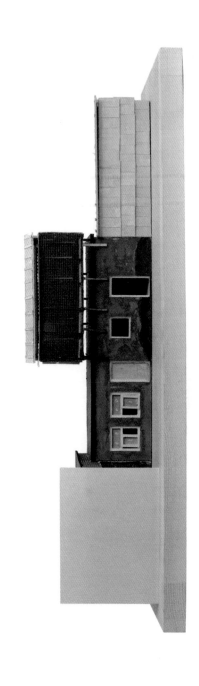

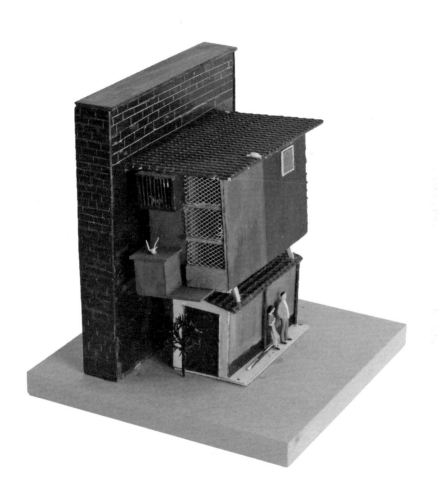

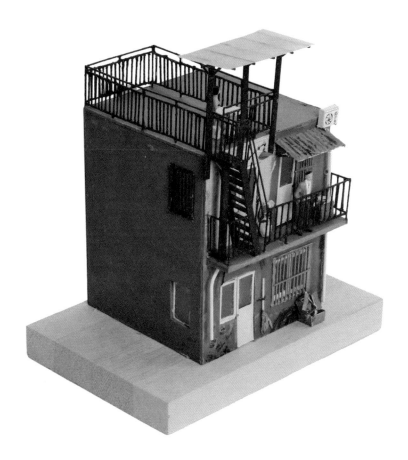

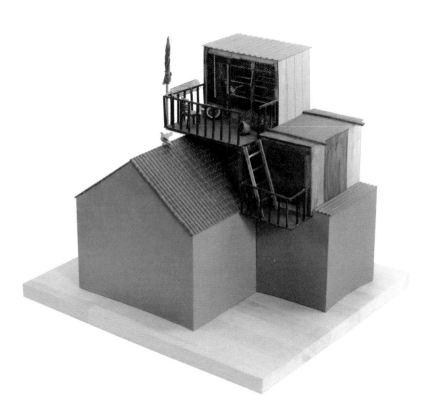

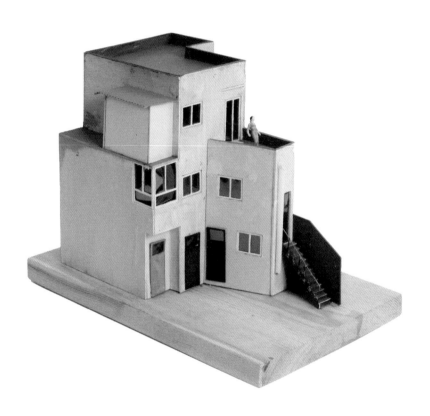

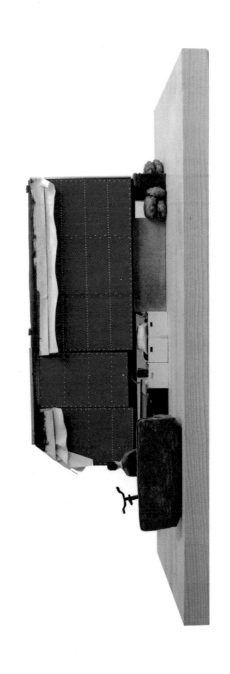

胡同蘑菇图画

Drawings of
Hutong Mushroom

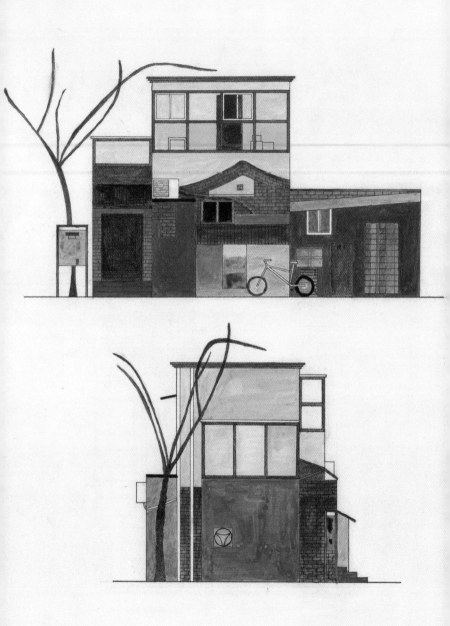

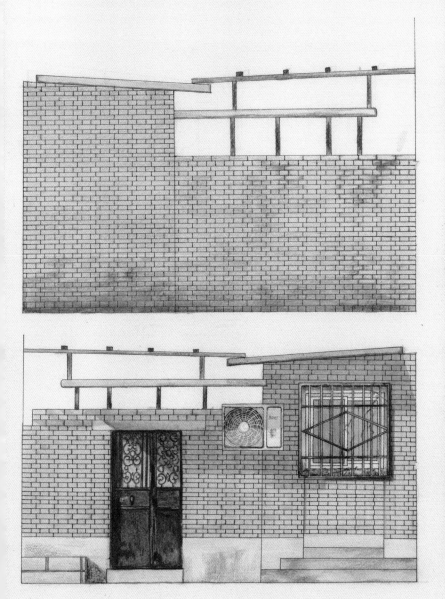

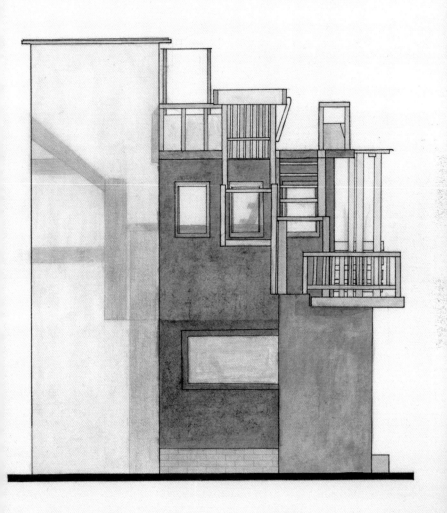

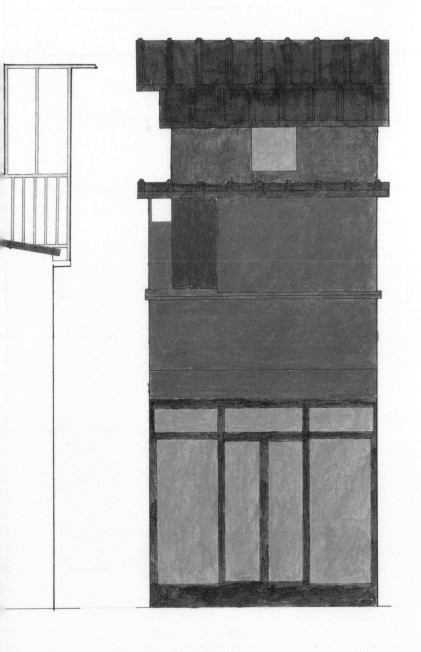

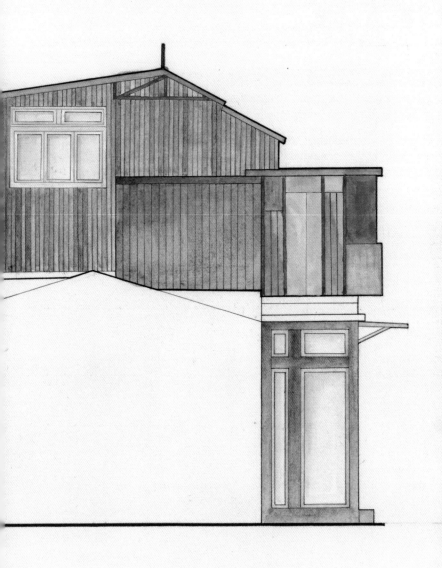

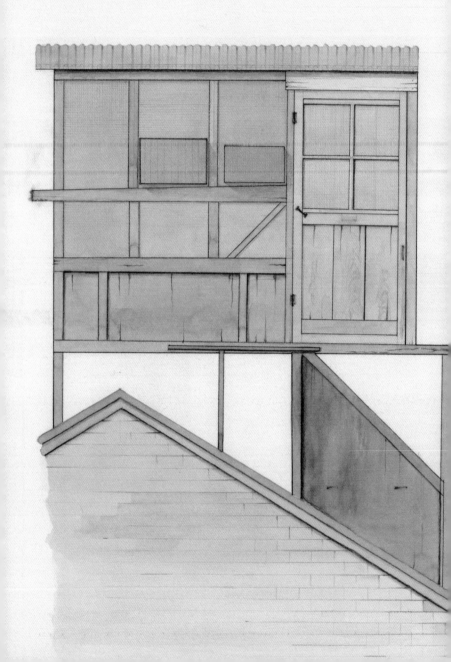

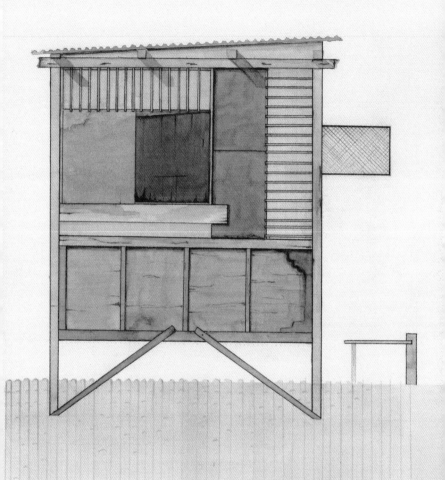

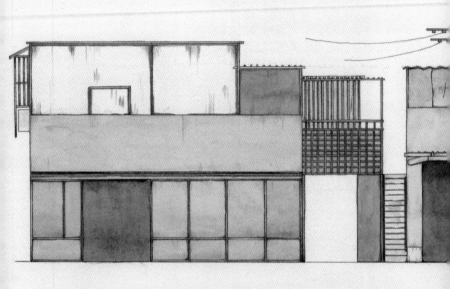

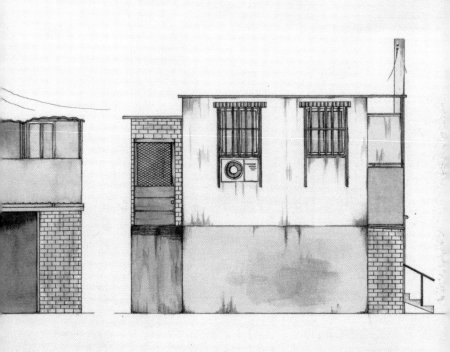

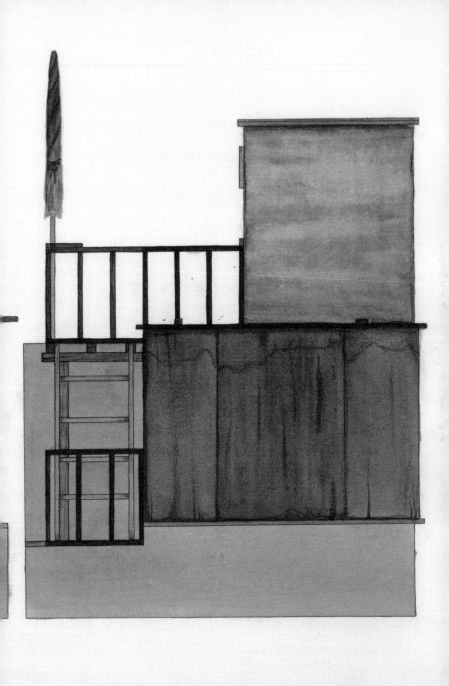

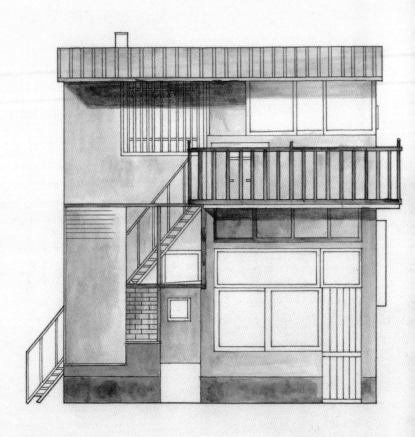

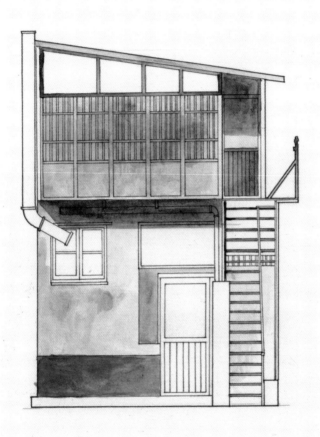

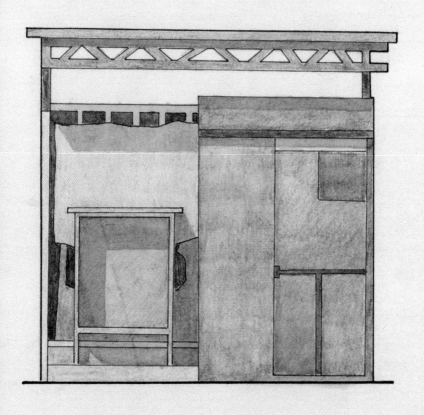

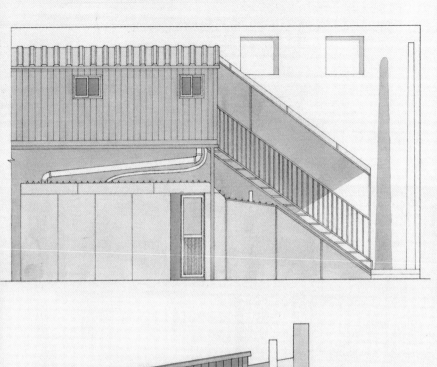

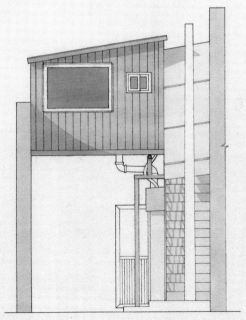

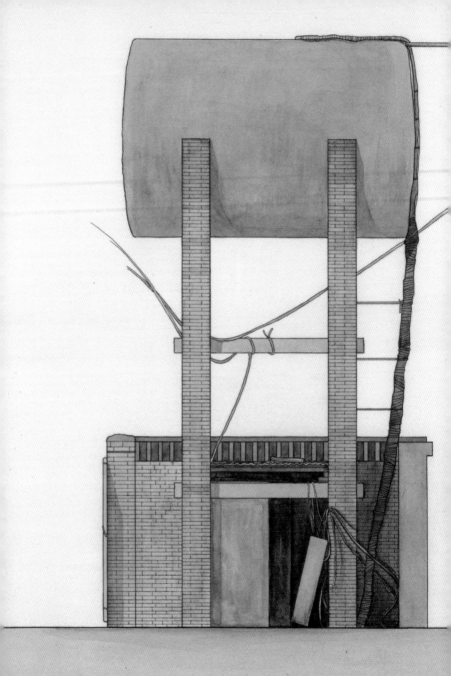

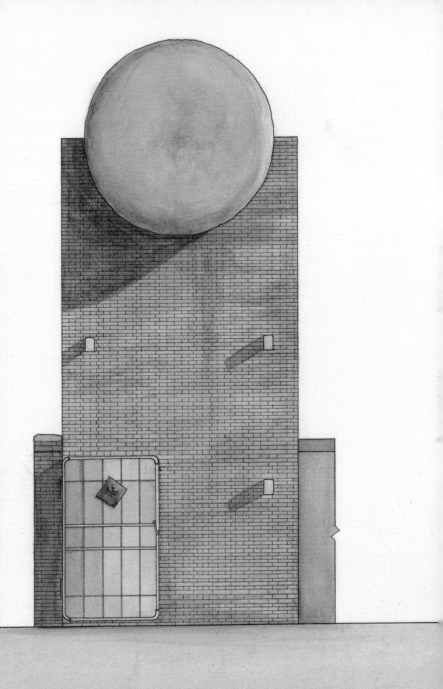

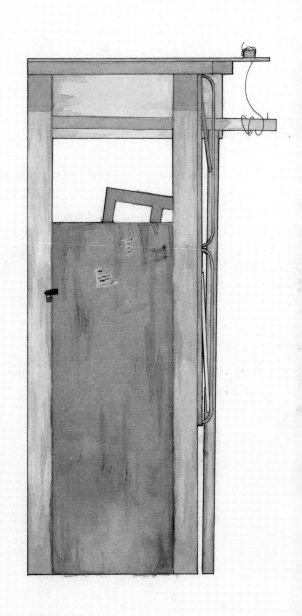

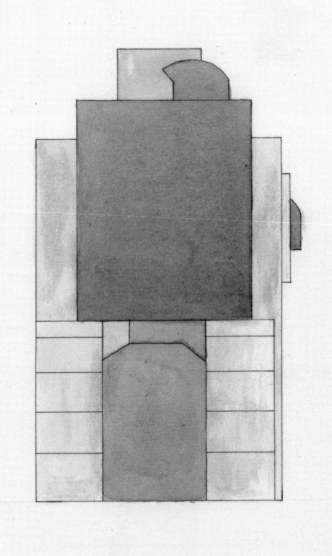

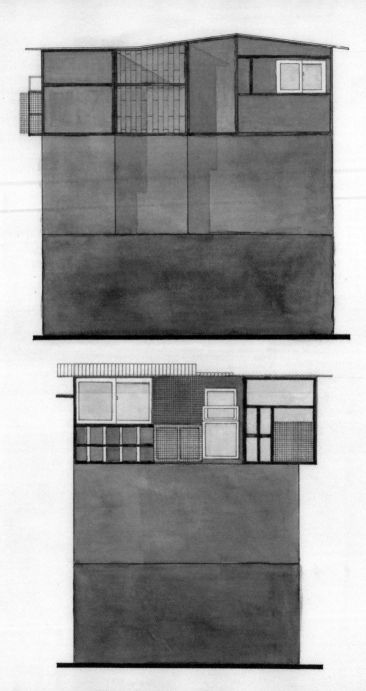

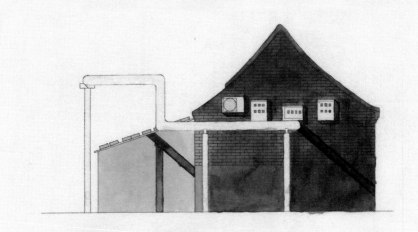

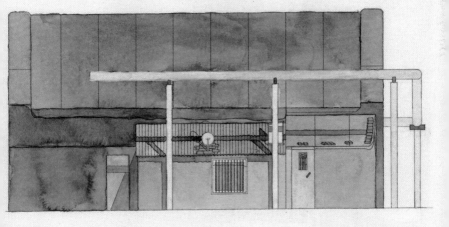

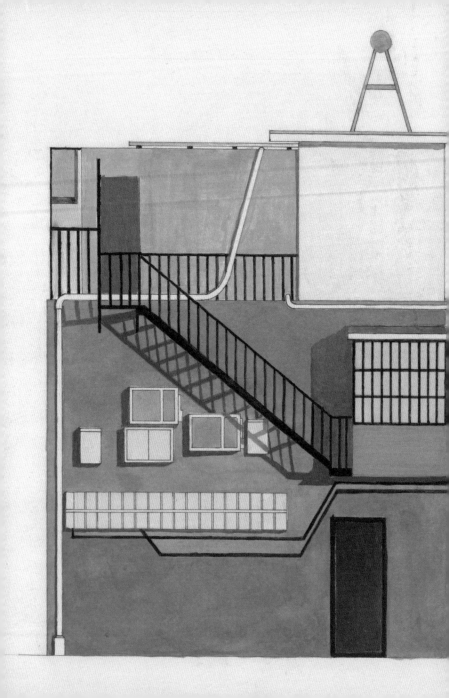

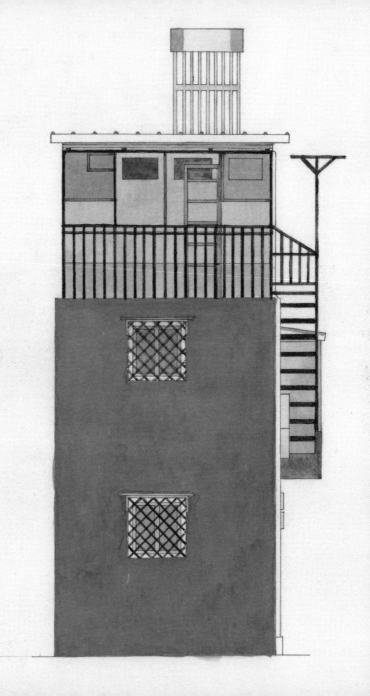

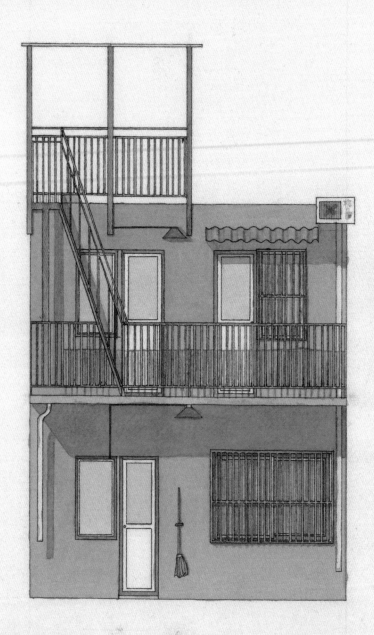

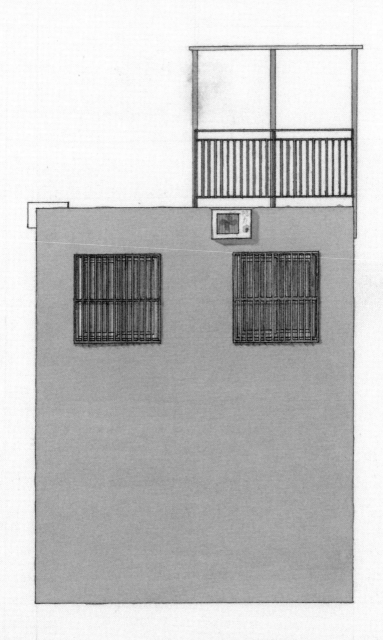

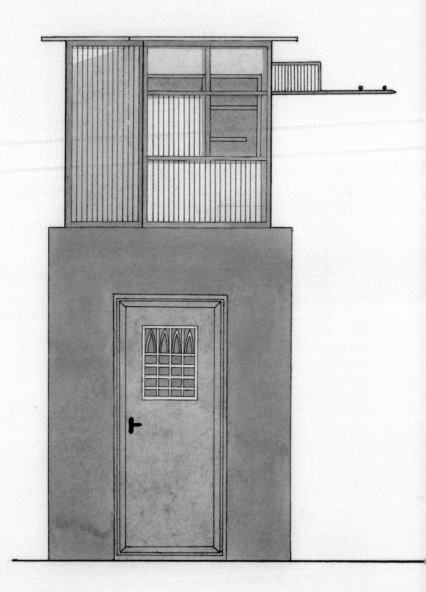

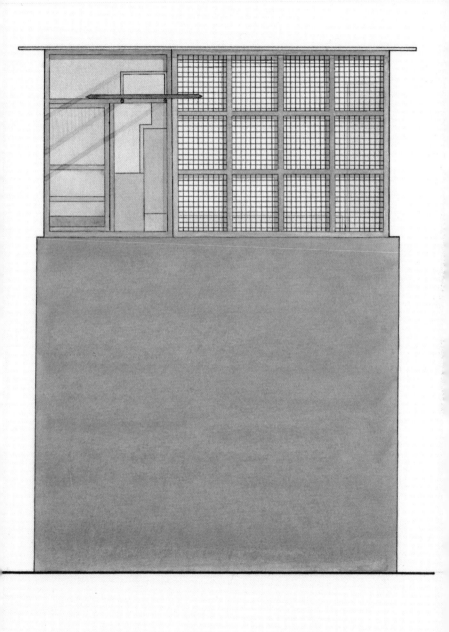

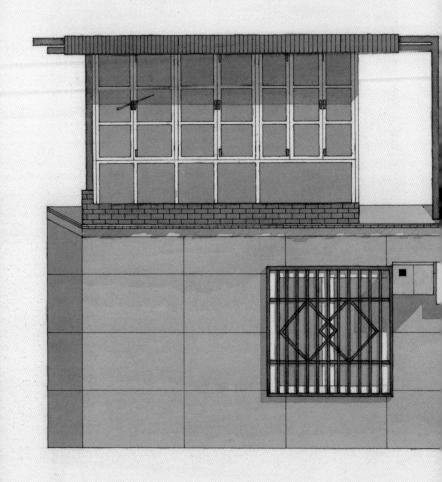

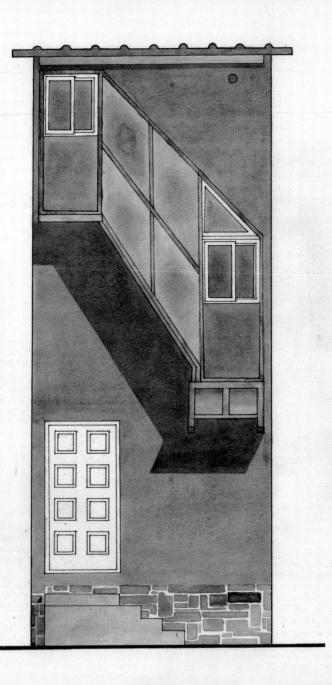

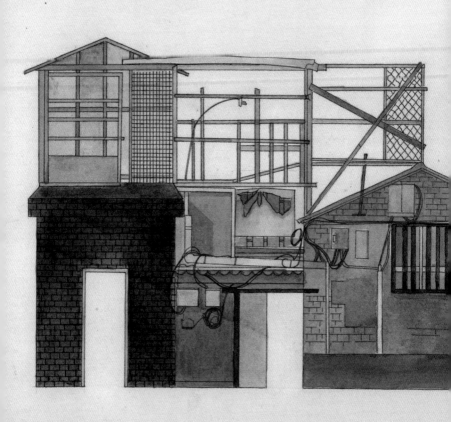

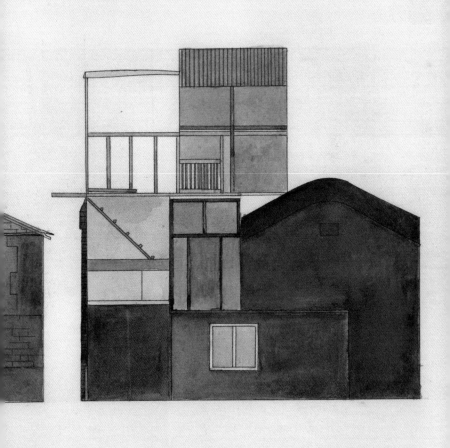

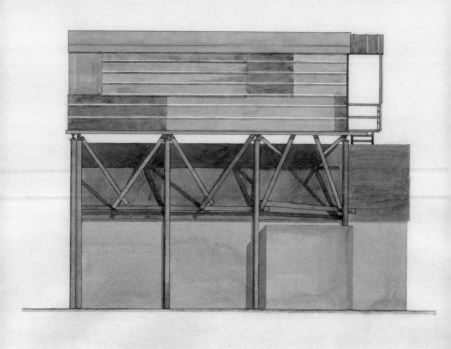
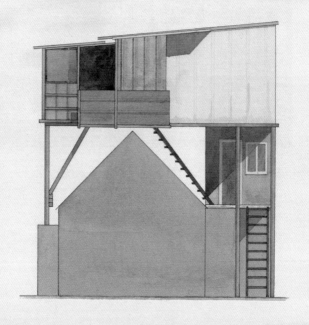

金秋野

胡同一课

胡同原本就是北京城本身,北京就是一条一条胡同沿着经线纬线织成的,把紫禁城织在中央;随着城市的扩张,胡同缩成一片灰蒙蒙的,好像老房子的窗纱,新的时候干干净净、透风透光,如今落满灰尘,孔隙都堵住了,城市随着它都变昏暗了。如今是现代的北京,摩天楼和宽阔的马路从周边升起,胡同依然在中间,好像脱离了时代似的。但它为北京增加了城市的厚度。在胡同为建筑师带来的全部教益中,有四个方面尤为重要,就是它形态上恬不知耻的坦诚、物料上捉襟见肘的可持续性、空间上迫不得已的丰富,以及整体结构上匪夷所思的复杂性。有这四点,明知它不可学,却还是要学一学。

1. 形态:无奈坦诚

胡同的形态是坦诚的。说"形态坦诚"是什么意思呢?大概是说,本来是怎样就是怎样,没有假装成别人的样子。说人坦诚往往是一种夸奖,但胡同的坦诚不是坦荡,不是坦然,而是坦白而不自知,不入褒贬,没有道德含义。胡同是城市里少有的"不自觉"的领域,没有被人们的企图心沾染,它维持原本的样子,是不得如此。一旦被关注,成了保护区,墙壁刷上厚厚的灰色涂料,就一点都不真实了。胡同的坦诚,在于它对当下的坦然接受,既不假装是别处的当下,也不假装是本地的往昔。就这样破旧着,任时光延宕,空气中飘来或浓或淡的油烟和泔水气味,水泥瓦耷拉下来用碎砖头压住,任梁柱倾圮随便撑一撑。胡同不假装是风光盛年,不掩饰自己的疲惫老态,它已经存在上千年了。胡同的样子就如同胡同里的居民——大门口的闲汉和蹲三轮的小贩,不在乎自己在别人眼里的样子,因此真实。

坦诚深刻是形式的最高追求。好比与人交往,我们希望对方的心灵纯净又深刻,不要浅薄又算计。如不能深刻,至少朴素不假装。形式亦然,假装成美式小镇的别墅区和假装成民国风貌的商业街之所以不好,是因为它们不真实。反映在专业语言上,是建造细节的缺失、建造逻辑的紊乱和视觉形式的虚假。然而,今天的城市,多多少少都是一种自觉不自觉的模仿,模仿的对象无所不包,就是没有本来的样子。在一片舞台布景般速成的、低分辨率的城市中,胡同的粗糙是高分辨率的,充满了真实的细节和时间的堆叠,哪怕这些细节并不美好,也与千篇一律的虚假城市风貌拉开了距离。然而这一切并不是主动追求的结果,胡同的真实乃因其不得不真实,它无可奈何地,甚至有点恬不知耻地坦诚着,在沉重的历史负担和居住压力下忍受基础设施的匮乏和物质形态的尴尬,依然我行我素。

2. 物料:极限拼贴

建造是物料组织的秩序。高级的造物像有机生命,自无机的世界中生出,有目的,有秩序,有独特的形态去与环境对话,最终又能降解、回归到环境中去。建造中,精巧性和集约性同等重要，将最低限度的材料最高效能地组织起来，本身就是美。从这个意义上讲，蛛网和贝壳比混凝土房子美妙得多。虽然不够精巧，胡同对物料的使用是高度集约的，它珍惜每一块砖头、每一根梁柱，认真表达着经济条件的制约，不去进行无谓的装饰。胡同居民的住宅没有石膏罗马柱，也不做大理石阳台。这里的建造有与生俱来的克制，哪怕简陋也在所不惜。

可循环特征，在建造中表现为物料组织的潜力和新旧拼贴的美学。依然是在物质条件的限制之下，胡同的建造者创造出意想不到的工作方式，各种材料、各种部件，无论新旧，都不假思索地并置在一起，让人瞠目结舌。红砖配翠绿的油漆木门；水泥瓦配生锈的金属网；坡屋顶上用钢柱撑起明黄色的活动板房，上面覆盖着蓝铁皮屋顶；废旧的木板围起封闭楼梯间，窗上猩红色的小遮雨棚，旧钢管焊出狭窄的楼梯，通往倾斜的阁楼……胡同提供了建筑形态的大百科全书，又因时间的浸润而透明。这里面充满了有意的误用：三轮车卸掉轮子成为铁皮板屋的支撑物；旧锅炉房改造成的蜗居，顶着刷了银粉的水箱；旧冰箱改成的置物架，旧地板条装饰的外立面和空调通风管编织的异形空间。几乎每一代房屋都有前代的余韵，残砖碎瓦和旧门窗扇以意想不到的方式拼贴进来，既不是手工艺传统的延续，也不受工业生产条件的制约，它只是极限状态下人对物质环境的急智，不假思索又无可奈何。

依然在不自觉中，胡同突破了建筑课本的局限，摸索出极限物质条件下的物料运用法则，既是集约的，也创造了意想不到的个体多样性。

3. 空间：变形自然

胡同的美，它自己不懂得欣赏，也不会被写进教科书。这是一种关系之美,来自于许多各具情态的多样个体,以彼此冲突纠结的方式堆积在一起,带着攻击性又不得不彼此退让,你来我往成为这样的姿态,好像一棵树上的无数片叶子,在争夺阳光的过程中相安于枝头,每片叶子对生存空间的索求组成了树冠的整体形象。如果说城市里的房屋是阵列中的士兵,迈着整齐的步伐前进,不必担心撞到一起,胡同里的房子就像是战场上的士兵,在混乱和喧嚣中向前冲锋,避免被弹片击中的同时寻找合适的掩体。这为胡同里的房子赋予额外的活力,它的造型不是建筑师在头脑数据库中搜肠刮肚的结果,而是自然赋予。

从控制论或自组织理论的层面,人们可以深究胡同面貌的自然形态的成因。形式,在胡同的建造过程中一定不是一个静态的菜单式变量,而是一种彼此角力的结果,每一栋房子都受到各种客观条件的制约,而呈现出"不得不如此"的面貌,这也让多余的造型思考不再必要。城里的房子,造型来自于专业性的思维过程,无论建筑师多么体贴、心思多么细腻、对多样性有怎样的追求,一个人的思维到底是有限的,"设计"出来的房子,形式上总有合法性的天然缺陷,而不能自如自在,人们不得不把这看作是"设计感"和静止的美学而接受下来,学着欣赏它,内心到底觉得不足。胡同的面貌尽管粗糙,丰富性却无与伦比,像无赖小儿,有城市不具备的泼辣顽皮,一种大大咧咧的生命力。尽管房屋在类型上基本一致,却没有任何形式上的雷同,相反每座建筑极尽所能地成为自我,从设计师的角度看未免是极出格的设计,事实上却不过是种种限制条件使然,这让我们对人造环境的丰富性多一层理解。当年筱原一男说"民居是蘑菇",大概就是这个意思吧,他的比喻多么形象!胡同是不是也可以看作某种聚落或原始民居呢？它只是刚好借鉴了现成的文明世界的建造法则。而在建造的知识归类上，它更接近于穴居。

在设计师眼里，胡同的形态对比出人的思维的不足，它的美是深层的、结构性的、关于组织方式的，与其粗糙的物质形态无关。胡同唤起了人们对自然的回忆，这里虽然并非一片树林，却有着类似于树林的形态特征：类型单一、组织精密、相互制约、循环永续。与人为设计的造型相比，胡同虽然带有明显的地域特征，本质上却并无风格，亦无"内涵"，人们无法对其进行象征层面的解读。这都是自然本身的特点。

自然到底是什么？是有机生命和无机物共同组成的静观世界，是一种万物共生共存、共同演化的方案，还是一套生命体竞争运行的机制？我们不得而知。在破解自然造物的密码之前，我们只能通过欣赏并读取它的外在形式，并思考如何接近那种复杂多样。毕竟吸引我们的不是它的外部特征，而是"相看两不厌"和彼此不相违的亲密感。胡同或许不能给人带来感官上的愉悦，但在机理上却超过了单调的城市设计，而带来"理智的愉悦"。

从这个意义上讲，胡同里即使没有一棵树，也是一种"自然态"的建造，带有自然造物的非线性特征。与原始村落和中世纪城镇一样，胡同像是自然假借人的双手创造的自然群落的变形，是人化的自然。

4. 复杂性：形式含义

胡同如蜂巢，它的物质形态是无数居民微观建造活动的总体效果。2010年北京内城人口密度高达23400人／平方公里，平摊在以单层居住模式为主的二环内老城区，像蚁穴。胡同建筑的建造模式是独特的,既不同于传统意义上的自发性建造,也与城市外围林立的商品住宅迥然有别。胡同的建造是包容性的,介于自发与人为规划之间。这是一种组织模式上的排中律,避免了城市空间构思的非黑即白,间接促成了城市面貌的自然态。在胡同建造之初,是有严整规划,并得到严格执行的。胡同的自然态是规则人为框架下千百年自发生长的结果,一种受控的失控状态:因能容纳,而成其大。

探讨胡同复杂性的成因,不是本文的目的所在。我们仅作最基本的推测。胡同在建造过程中,存在以下特点:(1)保持基本的空间尺度和交通结构大致不变。(2)强调"基本建造",从关系中生出形体,维持基本类型,坚持就地取材和经济上的最小化法则,严格从功能出发。(3)外在的均质和内在的多样性。可以说,胡同的形态是一种社会性的复杂"思维"的结果,以一种群体化和历时性的集体推力实现,进行细致的刻画和修改,它的建造过程非常不同于建立在现代规划设计基础上的简单线性城市构思。胡同的多样性对学校教育所提倡的现代设计体系提出疑问。与胡同的复杂形态相应的,是它在管理模式上的近乎无政府状态,即多元复杂的生活形态和高度简化的管理机制间的平衡。这也恰好与胡同"受控的失控"相呼应。

胡同的表观形态,相信很难引起审美人士的共鸣。那么,它吸引建筑师,或对现代建筑设计法则提出批评的方面到底是什么?我们如何评价一种形式,看它的准确程度?抽象程度?坦率程度?完整程度?社会效率?美学特征?经济效益?管理难度?我却认为,空间如同艺术作品,它的耐读性取决于其形式含义的展开方式,如果这件作品是个体性的,那么这种深度体现为它的文化厚度、它的言外之意、它的情感深度和象征性的准确性。如果它是集体性的,那么它的复杂性可以以自然来类比——整体简单,结构错综。错综是一种结构复杂性,它让我们对自然之美心有所感,不需要任何文明语言的介入,就可以在生命本能层面引起共鸣。与快速成型的现代城市相比,胡同的结构是一种跨越了时间和空间的拓扑结构或超级链接,每一栋房屋的缓慢演变都建立在原有房屋的基础上,同时与周边发生复杂生动、此消彼长的物质关联。这就从根本上有别于城里的房子,尽管后者也可能很复杂,但由于结构不是拓扑的,因此只具备简单的复杂性。这一点在城市道路设计上最为明显:立交桥越来越复杂,但因为通路不是拓扑形态,管道不能交叉,单元之间必须严格隔离,因此无论多么繁复都不能生出真正的复杂性,无法建立超级通路,只能受制于车流的潮汐。立交桥是复杂的,人们使用它,赞美它,却没有人站在桥底下流连忘返。

一个物质空间,哪怕完全是人造的,如果具有这种拓扑结构,也就可以说具备了自然的某些构造特征,表现在外部形态上,就能抚慰观察者的心与眼。胡同在物质上的低劣简陋,掩不住关系上的复杂多样。复杂性为形式赋予深度。

一段历史赋予城市一副面貌,我们熟悉的胡同并非向来如此,我们却以为它会一直如此。可是转瞬间,这样的胡同却要遁迹于历史深处了。在它欲走还留之际,我们不妨唤住它,让它给我们再上一课。

Jin Qiuye

A Lesson from Hutong

Beijing is all about Hutongs. It is woven by the stripes of Hutongs in the grid of latitudes and longitudes with the Forbidden City as the central node. As the city expanded, Hutongs shrank into a shade of grey like the window screens of the old houses. Clean and porous while they were new, they are now covered with dust and blocked by dirt, casting a shadow on the whole city. Today's Beijing is a modern city. Hutongs seem out of date with the surrounding arising skyscrapers and wide roads in contrast. However, Hutongs contribute to the rich layers of the city. Among the many enlightenments that Hutongs bring to architects, four of them are of great importance: the shameless honesty in form, the scarce sustainable material, the forced richness in space, and the bizarre complexity in structure. These four points of Hutongs make it hard to resist the urge for us to explore them further, regardless of any possible difficulties.

1. Form: Shameless Honesty

Hutongs are shameless in form. What is "shameless form"? It probably means the original look of an object without pretending to be something else. While honesty usually stands for a compliment to people, Hutongs' honesty is the frankness with no self-awareness and moral judgment. Hutongs are the rare "unconscious" area of the city. They have not been reached by ambitions and somehow have been forced to remain their original form. Once spotted and turned into the conservation districts, their genuineness would be destroyed simply by the thick paints on the wall. Hutongs' honesty lies in their acceptance of their presence, instead of the presence of others or the history of themselves. They bare the passage of time in their shabby clothes and carry on with the smell of fumes and swill blown in the air and drooped tiles and beams are covered with bricks and scraps. Hutongs do not fake the prime of their lives nor disguise their ageing, as they have existed for thousands of years. Hutongs and their residents are alike —— idlers outside the doors or hawkers on the tricycles. Their indifference to others' judgment makes them real.

Honesty and profundity are the highest pursuits in form. Just like making friends, we look for pure and deep hearts instead of superficiality and conspiracy. If profundity is not available, at least it could be simple and not pretending. It is the same for "form". Those villas faking American country houses and commercial streets of "Republic Era" style are bad because they are not real. Reflected in professional vocabulary, they are bad due to lack of construction details, chaos in construction logics, and falseness in visual form. However, our cities today are becoming imitators of almost anything but themselves, either with or without self-awareness. Compared to the city like a range of rapidly constructed stage sets in low resolution, the roughness of Hutongs is full of real details and stacks of time in high resolution. Although the details are not all pleasant, they are still able to keep Hutongs a distance from the false urban sceneries. But these do not result from an active chase but rather a compromise. It is without choice that Hutongs stay real, even shamelessly honest. They have to persist in their old ways and endure infrastructure shortage and awkwardness in physical form under the heavy historical burden and residential pressure.

2. Material: Super Collage

Construction is the order by which materials are organized. Advanced form is like an organic life born in the artificial world with a purpose, an order and a unique form. It talks to its environment and returns to the nature after being decomposed. In construction, delicacy and intensity are of the same importance. The organization of minimum material use in the highest efficiency is a beauty itself. From this perspective, spider web and shells are much better than concrete buildings. Although short of elegance, Hutongs are highly intensive in terms of material use. Each brick or column is treasured and used to sincerely express the constraint by economic conditions, without any meaningless decoration. The houses in Hutongs are not featured with Roman pillars or marble balconies. The construction in Hutongs is born to be restrained, even if the outcome is shabby.

In construction, sustainability is represented by the potential of material use and the aesthetics of the collages with old and new. Still, with limits imposed by physical conditions, the builders of Hutongs continuously offer their incredible creativity. All sorts of materials and components are spontaneously put together, which constantly amazes us: red bricks with a bright green wooden door, concrete tiles with rusty metal mesh, a yellow shelter on the sloping roof supported by steel columns and covered with blue iron panels, a closed staircase enclosed by the crappy wood boards with little scarlet canopies above the windows, a narrow staircase made of welded steel pipes that leads to the tilted attic, etc. Hutongs have offered us an encyclopedia of architectural forms that become transparent through the infiltration of time. Intentional misuse is everywhere: a tricycle with the wheels taken off becoming the supporting base for an iron panel house, a mini dwelling transformed from the old boiler room with a silver painted water tank on the top, shelves made of old refrigerators, façade decorated with the old floor tiles, and an irregular space woven by air conditioning vent. Almost every generation of the hutong houses has inherited some glamour from its ancestors. Broken tiles, old doors and windows are collaged in unexpected ways. They are neither the continuation of a handicraft tradition nor the product limited by industrial production. They are merely people's quick wits towards the material environment in some extreme conditions, both spontaneous and impotent.

Again, Hutongs have unconsciously broken the limitation in architecture textbooks and explored the theories of material application under extreme material conditions that are both intensive and unexpectedly individually diversified.

3. Space: Natural Transformation

The beauty of Hutongs would never be appreciated by themselves or be included in textbooks. It is a beauty of relationship that comes from various individuals. They are piled up and collide with each other with aggressiveness and compromise, interacting like the numerous leaves on the trees, which rest in peace next to each other while fighting for the sunlight. Every leaf's demand for the living space defines the overall profile of the tree. If buildings in the city are soldiers in the array that march in the same pace without having to worry about bumping into each other, the houses in Hutongs are like the soldiers in the field that bustle forward into the chaos, looking for cover while avoiding being shot at the same time. Therefore, the houses in Hutongs are extremely energetic with a style which is not a result from architects struggling to search in their mental database, but rather is given by the nature.

From the perspective of cybernetics or self-organization, we can dig deep to explore the causes of such a natural form of Hutongs. In the construction process of Hutongs, form is in no way a stable menu type variable but is the result of interactive competence. Each house has to obey the objective conditions and reveal its own "imperative" look, which makes it unnecessary to think twice about form or style. Houses in the cities owe their designs to the professional designers. But one man's thinking is limited anyway no matter how thoughtful, careful, or demanding he is about design diversity. "Designed" houses are born to have legitimate deficiencies and just cannot do as they want. And no matter how reluctant they are, people are trained to accept and appreciate them as the perfect examples of "design" and stationary aesthetics. Rough as they are, Hutongs' richness is incomparable. It is like a naughty boy with extreme vitality. Houses in Hutongs are basically identical in building type, however they share no similarities in building form. On the contrary, each building is trying its best to be itself. They might be outrageous designs to the architects, but in fact they are just the natural responses to their environmental limits, which helps us to understand their abundance and that of artificial living environments. That's probably what Kazuo Shinohara meant when he said "houses are mushrooms". What a vivid metaphor! Can we look at Hutongs as some type of settlement or original communities? They just happened to have borrowed some existing building techniques from the contemporary civilizations, although they are closer to caves in the classification of construction knowledge.

In designers' eyes, Hutongs' form surpasses the limits of human thinking. Their beauty is multi-layered, dealing with structure and arrangements and has nothing to do with their rough form and materials. Hutongs remind us of the nature. They are not forest but carry the same features such as single type, precise organization, mutual restraints, and sustainable recycling. Compared to manmade forms, Hutongs do not have a style or "meaning", although they have obvious geographical features. They are not metaphorical. All of these issues are the characteristics of the nature.

What is the nature? Is it the world of organic life and inorganic compositions? Is it an evolution program where everything coexist and develop? Or is it a set of competitive mechanisms followed by all the living things?

No one knows. Before decoding the secret of natural creation, we can only try to contemplate that complexity by reading and appreciating its external form. After all, what intrigues us is not its appearance but a feeling of closeness and mutual attraction. Hutongs may not bring us sensual pleasure, but they bring us a kind of "sensible pleasure" when their textures go well beyond the monotonous urban designs.

In this sense, Hutongs are the natural creations even without any trees inside because they have the nonlinear features of natural creation. Like the original villages and medieval towns, Hutongs look like the deformations of natural communities by hand——the man-made nature.

4. Complexity: Formal Meaning

Hutongs are like the honeycomb. The micro-construction activities of countless residents contribute to its material form. In 2010, the population density of the inner city of Beijing reached 23, 400 people / square kilometer, which means that people residing mostly in the one-storey houses within the 2nd ring road are living in a condition that's almost like the ants' nests. The construction model of Hutongs is unique which is different from both the traditional spontaneous construction and the commercial development outside the city. It is inclusive, and between spontaneity and artificial planning. It avoids the mistake of absolute judgment in urban space thinking; it is the law of excluded middle in organization, indirectly contributing to the natural appearance of the city. In the initial stage of its construction, it had a strategic plan and was carried out carefully. The natural state of Hutongs is the result of spontaneous growth under the principle frames defined by man. It is a controlled runaway state: becoming as comprehensive as it is inclusive.

The discussion of the causes of Hutongs' complexity is not our interest here. We only conduct some basic speculations. The construction process of Hutongs has the following characteristics: 1. Retained basic space scale and traffic flows structure in general; 2. "Basic construction", basic building types, forms out of relationships, local materials and minimal principles all strictly following functional needs; 3. External homogeneity and internal diversity. The form of Hutongs is the result of a complex "thinking" of sociality accomplished by the group and the collective thrust as time goes by. It is shaped in detail and modified with care, which makes it very different in construction from the linear urban planning thinking based on the modern city planning theory. The diversified Hutongs question the modern design theories advocated in architecture schools today. What comes with the diversity is Hutongs' anarchy in management. There is a balance between the plural and complex life there and the highly simplified management system, which echoes the "controlled runaway state" of Hutongs.

It is safe to say that Hutongs cannot appeal to an aesthete's taste by its form. Why do they attract architects? And what do they criticize in modern architecture design principles? How do we judge a type of form? From the degree of its precision? Abstraction? Honesty? Completeness? Social efficiency? Aesthetic feature? Economic benefits? Management difficulty? I think that space is like artwork, its readability depends on the way it displays its meaning. The deep meaning of an individual work is embedded in the cultural thickness,

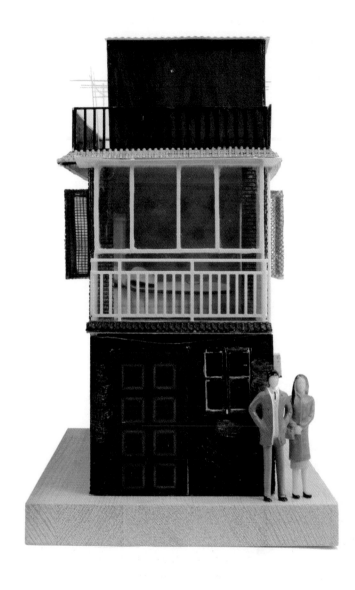

connotation, emotional depth and precision in metaphor. The complexity of a collective group is analogical to the nature, structurally intricate but holistically simple. Intricacy refers to the structural complexity. It arouses in our body a confusion of the nature's beauty on the intuitive level without any languages. Compared with the rapidly developed modern city, the structure of Hutongs is a topology and hyperlink that crosses the border of time and space. The slow evolution of each house has a vivid, complicated and competitive relation with its surroundings based on its original foundation, which distinguishes itself from the buildings in the city. The structure of the latter is not topological, it only possesses a "simple complexity" even though it can be very complex from its own side. The urban road planning is the best example: the overpasses are becoming increasingly complex, but they are not generated in the topological way, meaning that the ducts cannot intersect and units must be strictly separated from each other with no real complexity or unimpeded roads. Overpasses are complex, people use them and admire them, but no one would wander under them.

With such typological structure, a material space consists of the nature's structural quality, even if it is completely man-made. Its appearance can please the observer's eyes and mind. The poor and shabby materials used in Hutongs cannot erase its complexity in relationships. Complexity grants depth to form.

Every period of time in history leaves marks on its cities. While we think they will last, the Hutongs we know today are no longer those of the past. Thye escape in a second. While they are still hesitating, we might as well hold they back for another lesson.

研究团队

Research Team

研究人员 Researchers

赵相如 Zhao Xiangru

尹蔚然 Yin Weiran

杨文星 Yang Wenxing

薛苏洪 Xue Suhong

谢佳羽 Xie Jiayu

肖梦 Xiao Meng

王逸品 Wang Yipin

乔大漠 Qiao Damo

马雨菲 Ma Yufei

吕后晨 Lyu Houchen

刘俪婷 Liu Liting

栗紫薇 Li Ziwei

李博文 Li Bowen

郎颖晨 Lang Yingchen

焦天 Jiao Tian

陈宇 Chen Yu

陈佳怡 Chen Jiayi

曹乐行 Cao Lexing

蔡斯巍 Cai Siwei

郭嘉月 Guo Jiayue

牛倩茹 Niu Qianru

王炜 Wang Wei

李俏然 Li Qiaoran

刘贺轩 Liu Hexuan

张雅薇 Zhang Yawei

马彪 Ma Biao

李炳华 Li Binghua

薄萌钰 Bo Mengyu

方玥莹 Fang Yueying

郝好 Hao Hao

陈肇晖 Chen Zhaohui

刘铭韬 Liu Mingtao

渠畅 Qu Chang   郭佳玉 Guo Jiayu

胡蓝 Hu Lan   张茜越 Zhang Xiyue

龙大宇 Long Dayu   吴天博 Wu Tianbo

吴曦 Wu Xi   徐可薇 Xu Kewei

侯明慧 Hou Minghui   邢璐阳 Xing Luyang

闫梦瑶 Yan Mengyao

研究主持 Research Directors

金秋野 Jin Qiuye   李涵 Li Han

指导教师 Instructors

王韬 Wang Tao   徐跃家 Xu Yuejia

刘烨 Liu Ye   齐莹 Qi Ying

郝石盟 Hao Shimeng   郭龙 Guo Long

程艳春 Cheng Yanchun 张振威 Zhang Zhenwei

李煜 Li Yu   刘璁 Liu Cong

余浩 Yu Hao   王晶 Wang Jing

孙立 Sun Li

杨震 Yang Zhen

钟真 Zhong Zhen

阚玉德 Kan Yude

韩风 Han Feng

摄影 Photographers

沈群松 Shen Qunsong

张冠峰 Zhang Guanfeng

翻译 Translators

韩建屋 Han Jianwu

崔中鼎 Cui Zhongding

杨梦溪 Yang Mengxi

陈鹏宇 Chen Pengyu

Jared Baker

作者简介

About the Authors

李涵,绘造社主持人,国家一级注册建筑师。毕业于中央美术学院和皇家墨尔本理工大学。创作领域包括建筑设计、建筑绘画和城市研究。著有城市绘本《一点儿北京》和《一点儿北京·大栅栏》。译有《新兴建构图集》。

Li Han, principal of Drawing Architecture Studio, National Class 1 Registered Architect in China. He received B. Arch. from Central Academy of Fine Arts in Beijing, China and M. Arch. from RMIT University in Melbourne, Australia. His practices include architectural design, architectural drawing and urban studies. He is the author of graphic novels *A Little Bit of Beijing* and *A Little Bit of Beijing · Dashilar* and the translator for *Atlas of Novel Tectonics* by Reiser+Umemoto.

金秋野,清华大学博士,中国建筑学会理事,北京建筑大学教授,设计基础部主任,建筑评论研究所主持人,学者和建筑评论家。《建筑师》和《城市设计》编委,《建筑学报》特约学术主持。研究领域包括面向当代本土建筑实践的建筑评论;传统设计语言的现代转化;当代建筑思想脉络等。著有《尺规理想国》、《异物感》等评论集及《乌有园》等系列出版物,也是《光辉城市》、《透明性》等理论专著的译者。

Jin Qiuye, Ph. D. of Tsinghua University, member of the Architectural Society of China, professor of Beijing University of Civil Engineering and Architecture, architectural critic. Host of Architectural Criticism Research Institute. Research fields include the translation of the Chinese traditional design language into modern architecture and the contemporary architectural thoughts in China. He is the author of the critic essay collection *Utopia on the Drawing Board and Arcadia: Painting and Garden*. His translation includes *Radiate City* of Le Corbusier and *Transparency* of Colin Rowe.

图书在版编目（CIP）数据

胡同蘑菇 / 李涵，金秋野著. —北京：中国建筑工业出版社，2018.5

（零度城市研究系列）

ISBN 978-7-112-21917-9

Ⅰ.①胡… Ⅱ.①李… ②金… Ⅲ.①建筑设计—研究—北京—汉、英 Ⅳ.①TU2

中国版本图书馆CIP数据核字（2018）第044858号

本项目由"北京未来城市设计高精尖创新中心——城市设计理论方法体系研究"资助，项目编号UDC2016010100

本项目由教育部人文社会科学研究一般项目"自然态建造思想研究"资助，项目编号16YJAZH024

责任编辑：戚琳琳　李　婧
责任校对：卢欣甜
书籍设计：康　羽

零度城市研究系列
Urban Studies Degree Zero Series
胡同蘑菇
Hutong Mushroom
李　涵　金秋野　著
Li Han　Jin Qiuye
*
中国建筑工业出版社出版、发行（北京海淀三里河路9号）
各地新华书店、建筑书店经销
北京点击世代文化传媒有限公司制版
北京富诚彩色印刷有限公司印刷
*
开本：850×1168毫米　1/32　印张：5½　字数：132千字
2018年10月第一版　2018年10月第一次印刷
定价：68.00元
ISBN 978-7-112-21917-9
（31660）

版权所有　翻印必究
如有印装质量问题，可寄本社退换
（邮政编码 100037）